RECOMMENDED FOR HIGH SCHOOL AND UNIVERSITY STUDENTS.

The Big Bang: The Amazing Story of How Science Unveils the Secrets of the Universe

RYBAIEV VOLODYMYR

The Big Bang: The Amazing Story of How Science Unveils the Secrets of the Universe

The Big Bang: The Amazing Story of How Science Unveils the Secrets of the Universe

Table of contents

Introduction 4

History of Big Bang Research 6

Basic Characteristics and Stages of the Big Bang, from Singularity to Inflation and Recombination. 12

    Singularity 12

    Planck Epoch 13

    Grand Unification Epoch 14

    Electroweak Interaction Epoch 14

    Quark Epoch 16

    Hadron Epoch 17

    Lepton Epoch 18

    Photon Epoch 19

Traces and evidence of the Big Bang, such as cosmic microwave background radiation, abundance of light elements, and the expansion of the Universe. 22

    Cosmic Microwave Background Radiation 22

    Abundance of light elements 24

    Expansion of the Universe 25

Riddles and Problems of the Big Bang, such as dark matter, dark energy, antimatter, and the horizon problem. 30

    Dark Matter 30

    Dark Energy 32

    Antimatter 34

    The Horizon Problem 37

Alternative and extended models of the Big Bang, such as the multiverse, cyclic universe, quantum gravity, and string theory. 40

    The Multiverse 40

    The Cyclical Universe 43

    Quantum Gravity 44

What we know and don't know about the Big Bang and the birth of the Universe, and the prospects for further research. 47

Introduction

The universe is one of the most astonishing and mysterious realities we encounter in our lives. Where did it come from? How is it structured? What is its fate? These questions have occupied the minds of people from ancient times to the present day. Modern science offers us one of the most compelling and well-founded answers to these questions: the theory of the Big Bang.

The Big Bang is a scientific model that describes the birth and development of the universe from an extremely hot and dense state of matter and energy that existed approximately 13.8 billion years ago. According to this model, the universe began its existence with a massive explosion that caused its rapid expansion and cooling. During this expansion, the first elementary particles, atoms, stars, galaxies, and other structures that we observe today formed.

The theory of the Big Bang is not a simple hypothesis or guess. It is based on a wealth of experimental data and theoretical calculations that confirm its correctness and accuracy. However, this does not mean that the theory of the Big Bang is without problems or does not raise questions. On the contrary, the more we learn about the universe, the more we encounter new mysteries and paradoxes that require further study and explanation.

In this book, we will explore the key ideas and facts of the Big Bang theory, as well as its history and prospects. We will learn how people came to understand the origin of the universe, what experiments and observations confirm the theory of the Big Bang, what stages and characteristics the

Big Bang had, what traces it left in our universe, what problems and mysteries it generates, and what alternative and extended models it stimulates. We hope that this book will help you better understand our universe and your place in it.

History of Big Bang Research

The Big Bang is not just a scientific theory but also the result of a long and fascinating journey of the human mind in search of answers about the origin and fate of the universe. This journey began in ancient times when people pondered philosophical and religious questions about what existed before time and space began, and how the world we see around us came into being. These musings gave rise to numerous myths and legends that attempted to explain the origin of the universe through divine creation, a cosmic egg, a giant tree, or other symbols.

However, a real breakthrough in understanding the universe occurred in the early 20th century when two revolutionary scientific theories were developed: Albert Einstein's general theory of relativity and quantum mechanics. These theories allowed for the description of physical laws that operate in the universe on all scales, from the largest to the smallest. They also opened new horizons for exploring space through astronomical observations and experiments.

In 1916, Einstein published his work "The Foundations of General Relativity," in which he completed the development of the relativistic theory of gravity. This theory showed that gravity is not a force but a consequence of the curvature of spacetime caused by matter and energy. This meant that spacetime is not static and absolute but can change under the influence of various factors. It also meant that the universe does not necessarily have to be stationary and homogeneous, as previously believed.

In 1922, Russian physicist and mathematician Alexander Friedmann, based on Einstein's equations, showed that the universe could be dynamic and expand or contract. He also proposed several models of the universe with different parameters. One of these models corresponded to the observed universe and predicted its beginning from a state of infinite density and temperature.

In 1927, Belgian priest and astronomer Georges Lemaître independently derived a model of an expanding universe and suggested that it originated from a "primeval atom" - a singularity where all matter and energy were concentrated. He also estimated the age of the universe to be around 10-20 billion years.

In 1929, American astronomer Edwin Hubble discovered that distant galaxies are receding from us at speeds proportional to their distances. This was the first experimental confirmation of the theory of the expanding universe. Hubble also determined the Hubble constant, which is the proportionality coefficient between the velocity of galaxy recession and their distance. This constant allows for the estimation of the age and size of the universe.

In 1931, Einstein acknowledged the correctness of Lemaître and Friedmann's work and abandoned his cosmological constant, which he introduced into his equations to obtain a stationary model of the universe. He called it his "greatest mistake." However, it was later found that the cosmological constant could have physical significance and describe the effects of dark energy.

In 1948, Russian physicist George Gamow and his students Ralph Alpher and Robert Herman proposed the theory of primordial nucleosynthesis - the process of forming light elements (hydrogen, helium, lithium) from the primordial plasma in the early universe. They also predicted the existence of relic radiation - the residual thermal radiation that fills the universe after its cooling and recombination.

In 1964, American radio astronomers Arno Penzias and Robert Wilson accidentally discovered relic radiation in the form of microwave noise emanating from all directions. They were awarded the Nobel Prize in Physics in 1978 for this discovery. Relic radiation became one of the most important pieces of evidence for the Big Bang theory.

In the 1980s and 1990s, new modifications of the Big Bang theory were developed, such as inflationary cosmology, quantum cosmology, string theory, and the multiverse theory. These models attempt to explain some of the mysteries and problems of the Big Bang theory, such as the flatness of the universe, its homogeneity and isotropy, the horizon problem, the monopole problem, the problem of the initial singularity, and others.

In the 1990s and 2000s, intensive space exploration continued using various space observatories and telescopes operating in different ranges of the electromagnetic spectrum. Space observatories are telescopes or other instruments located in space, many of which are in orbit around Earth. They have several advantages over ground-based observatories, such as the absence of atmospheric distortion and interference, the ability to observe objects in ranges that do not penetrate the atmosphere, and greater sensitivity and resolution.

Among the most well-known and important space observatories are:

The Hubble Space Telescope: Launched in 1990 by NASA and ESA, it operates in the visible, ultraviolet, and near-infrared ranges. It has made numerous discoveries and captured images of cosmic objects such as galaxies, stars, planets, nebulae, and black holes. It has also measured the distances to stars and galaxies, determined the age of the universe and the Hubble constant, discovered the acceleration of the universe's expansion, and dark energy.

The Chandra X-ray Observatory: Launched in 1999 by NASA, it operates in the X-ray range. It studies high-energy phenomena in space, such as supernovae, neutron stars, pulsars, black holes, active galactic nuclei, and hot interstellar gas.

The Spitzer Space Telescope: Launched in 2000 by NASA, it operates in the far-infrared range. It studies cold and dusty objects in space, such as star-forming clouds, protoplanetary disks, exoplanets, and distant galaxies.

The Planck Observatory: Launched in 2009 by ESA, it operates in the microwave and submillimeter ranges. It measures the anisotropies of the cosmic microwave background radiation with high precision and resolution, creates maps of polarization and radiation spectrum. It also observes other cosmic microwave sources, such as stars, galaxies, and galaxy clusters.

The Hubble, Chandra, and Spitzer observatories are part of NASA's "Great Observatories" program, which aims to study the universe in different ranges of the electromagnetic spectrum. In addition to them, the program includes the Compton Observatory, which operated in the gamma-ray range from 1991 to 2000.

In addition to these observatories, there are other space telescopes that operate in different ranges of the electromagnetic spectrum and study various aspects of space. For example, the Integral Observatory (ESA) operates in the gamma and X-ray ranges, the Herschel Observatory (ESA) operates in the submillimeter and far-infrared ranges, and the Gaia Observatory (ESA) operates in the visible range and creates a three-dimensional map of our Galaxy.

Currently, efforts are underway to create new space observatories and telescopes that will deepen our understanding of the universe and open up new horizons for astronomy. Among them, we can mention:

The James Webb Space Telescope: Planned for launch in 2021 by NASA, ESA, and CSA, it will operate in the near and mid-infrared ranges. It will succeed the Hubble Space Telescope and will feature a 6.5-meter diameter mirror (compared to Hubble's 2.4 meters). It will be capable of observing the most distant and faint objects in space, such as the first stars and galaxies, as well as exoplanets and protoplanetary disks.

The Advanced Telescope for High ENergy Astrophysics (Athena): Planned for launch in 2031 by ESA, it will operate in the X-ray range. It will become the most powerful X-ray telescope in history and will be capable of studying the structure and evolution of the hot universe, including black holes, galaxy clusters, active galactic nuclei, and other high-energy phenomena.

The Laser Interferometer Space Antenna (LISA): Planned for launch in 2034 by ESA, it will operate in the gravitational wave range. It will be the first space-based gravitational wave detector and will consist of three satellites forming an equilateral triangle with sides of approximately 2.5 million kilometers. It will be able to detect gravitational waves from the merger of black holes, neutron stars, white dwarfs, and other cosmic objects.

The Spectrum-Roentgen-Gamma (Spektr-RG) Telescope: Launched in 2019 by Roscosmos and DLR, it operates in the X-ray range. It consists of two telescopes: "ART-XC" (Russian) and "eROSITA" (German). It is designed to create a complete map of the X-ray sky and study dark matter, dark energy, black holes, and galaxy clusters.

In this chapter, we have discussed the development of the study of the Big Bang from the early 20th century to the present day. We have learned about the scientific theories and experimental discoveries that form the basis of this theory, as well as the space observatories and telescopes used to confirm and refine it. In the next chapter, we will examine the key characteristics and stages of the Big Bang, from singularity to inflation and recombination.

Basic Characteristics and Stages of the Big Bang, from Singularity to Inflation and Recombination.

In this chapter, we will examine what happened in the universe from the moment it emerged from a point of infinite density and temperature to the formation of the first atoms and molecules. We will learn about the physical laws and processes that shaped the evolution of the universe during different periods of its history, as well as the types of matter and radiation that dominated it. We will also explore some theoretical assumptions and hypotheses that attempt to explain certain mysterious aspects of the Big Bang.

Singularity

According to the standard model of the Big Bang, the universe originated from a state of singularity, which is a point of infinite density, temperature, and spacetime curvature. In this point, none of the known physical laws applied, so we cannot describe or understand what happened in it or before it. We can only speculate that the singularity resulted from the collapse of a previous universe or arose from nothing due to quantum fluctuations.

At some point in time, which we denote as $t=0$, the singularity began to rapidly expand and cool. This moment is known as the Big Bang. However, it wasn't an explosion in the conventional sense, as there was no center or direction of expansion. The universe was not expanding into a larger space; instead, spacetime itself was stretching and creating new regions. There was no sound or light either, as there was no matter or radiation present.

Planck Epoch

The first phase of the universe's expansion lasted only about 10^{-43} seconds and is called the Planck Epoch. It is the earliest period in the history of the universe that we can describe using our physics. During this stage, the temperature of the universe was around 10^{32} Kelvin, and the density was about 10^{94} grams per cubic centimeter. These values were so high that all four fundamental forces - gravity, electromagnetism, strong, and weak forces - were unified into a single fundamental interaction. Matter and radiation didn't exist as separate entities but were mixed together in a quantum foam where virtual particles and antiparticles constantly appeared and disappeared.

The Planck Epoch represents the limit of applicability of our modern physics, as describing it fully requires a theory of quantum gravity, which doesn't yet exist. Therefore, we cannot precisely say what occurred during this period and what mechanisms governed the behavior of the universe. One hypothesis suggests that inflation occurred during the Planck Epoch—an exponential acceleration of the universe's expansion that stretched its size by many orders of magnitude in a very short time. Inflation could have resolved some problems of the standard Big Bang model, such as the horizon problem, the flatness problem, and the magnetic monopole problem. However, inflation itself requires explanation as it is unknown what could have triggered its beginning and end.

Grand Unification Epoch

After the Planck Epoch, the grand unification epoch began, lasting for about 10^{-36} seconds. During this stage, the temperature of the universe dropped to around 10^{28} Kelvin, and the density decreased to about 10^{76} grams per cubic centimeter. As a result, the fundamental interactions separated, with gravity becoming an independent force distinct from the other three. However, the electromagnetic, strong, and weak interactions were still unified into a single electroweak interaction. Matter and radiation continued to exist in a quantum foam state, but the first elementary particles, such as quarks and leptons, started to form.

The grand unification epoch is also a subject of speculation and hypotheses because describing it requires a theory of quantum gravity. One possible theory attempting to unify all four fundamental forces is called string theory. According to this theory, all elementary particles are manifestations of a tiny one-dimensional string that can vibrate in different modes. It is also postulated that spacetime has more than four dimensions (up to 11), but the additional dimensions are compactified into extremely small sizes and are not observable.

Electroweak Interaction Epoch

The next phase of the universe's expansion is called the electroweak interaction epoch, which lasted for about 10^{-12} seconds. During this stage, the temperature of the universe dropped to around 10^{15} Kelvin, and the density decreased to about 10^{40} grams per cubic centimeter. As a result, the strong interaction separated from the electroweak interaction and became an independent force. However, the electromagnetic and weak interactions were still unified into a single electroweak interaction. Matter and radiation continued to form from the quantum foam but became more stable and diverse. Quarks began to combine to form hadrons such as protons and neutrons, while leptons started to form neutrinos and electrons.

The electroweak interaction epoch is the subject of study in modern high-energy physics, as its description requires the Standard Model theory. According to this theory, all matter particles are manifestations of two types of fermions: quarks and leptons. Quarks have a strong charge and can form composite particles called hadrons. Leptons do not have a strong charge and can exist independently. All matter particles have their antiparticles, which have opposite charges and other characteristics. All four fundamental interactions are mediated by bosons, which are massless particles or particles with spin 1. The strong interaction is mediated by gluons, the electromagnetic interaction by photons, the weak interaction by W and Z bosons, and gravity by gravitons.

One of the main problems of the Standard Model is that it cannot explain the origin of particle masses. For this purpose, the Higgs theory was proposed, which suggests the existence of a special field called the Higgs field that permeates all of spacetime and interacts with particles, giving them mass. The more a particle interacts with the Higgs field, the greater its mass. The Higgs field also has its own particle, the Higgs boson, which was experimentally discovered in 2012 at the Large Hadron Collider.

Quark Epoch

After the electroweak interaction epoch, the quark epoch began, lasting for about 10^{-6} seconds. During this stage, the temperature of the universe dropped to around 10^{13} Kelvin, and the density decreased to about 10^{20} grams per cubic centimeter. As a result, the electromagnetic and weak interactions separated from each other and became independent forces. Additionally, a violation of symmetry between matter and antimatter occurred, meaning there were more particles than antiparticles in the universe. This phenomenon is known as baryogenesis and still lacks a satisfactory explanation. Matter and radiation continued to form from the quantum foam but began to dominate over it. Quarks and leptons became the primary constituents of matter, but quarks were not yet able to form hadrons due to the high temperature. Instead, they formed a plasma of free quarks and gluons called quark-gluon plasma.

The quark epoch is the subject of study in modern high-energy physics, as its description requires the theory of quantum chromodynamics. According to this theory, quarks have a color charge that can take six values: red, green, blue, antired, antigreen, and antiblue. Gluons also have a color charge, which is a combination of two colors or anticolors. The strong interaction between quarks and gluons is mediated by the exchange of gluons, which can change the color charge of the particles. However, the strong interaction has a special property called asymptotic freedom, which means that at high temperatures or energies, the strength of the interaction between quarks and gluons decreases, while at low temperatures or energies, it increases. This leads to the phenomenon that at high temperatures, quarks and gluons can freely move within the plasma, while at low temperatures, they are confined into composite particles called hadrons.

Hadron Epoch

After the quark epoch, the hadron epoch began, which lasted for about 1 second. During this stage, the temperature of the universe dropped to 10^{10} Kelvin, and the density decreased to 10^{11} grams per cubic centimeter. As a result, the first composite particles made of quarks, called hadrons, formed. Hadrons are divided into two classes: baryons, which consist of three quarks, and mesons, which consist of a quark and an antiquark. Examples of baryons include protons and neutrons, while examples of mesons include pions and kaons. In the hadron epoch, intense annihilations occurred between particles and antiparticles, resulting in the conversion of a large portion of matter into radiation. However, due to a violation of symmetry between matter and antimatter in the electroweak interaction epoch, a small excess of matter over antimatter remained, amounting to about one particle per 10 billion antiparticles. This excess of matter is what formed all the visible matter in our Universe.

The hadron epoch is the subject of study in modern high-energy physics, as its description requires the theory of quantum chromodynamics. According to this theory, quarks and gluons obey the law of conservation of color charge, meaning the total color charge of a system of particles remains unchanged during interactions. This leads to the fact that quarks and gluons cannot exist in a free state but only as constituents of colorless combinations known as hadrons. This phenomenon is called color confinement. There is also a phenomenon known as asymptotic freedom, which means that at high temperatures or energies, the strength of the interaction between quarks and gluons decreases, while at low temperatures or energies, it increases.

Lepton Epoch

After the hadron epoch, the lepton epoch began, lasting for about 10 seconds. During this stage, the temperature of the Universe dropped to 10^9 Kelvin, and the density decreased to 10^9 grams per cubic centimeter. As a result, the first stable atomic nuclei, consisting of protons and neutrons, formed. However, the temperature was still too high for the formation of atoms because electrons were free and unable to bind with nuclei. Therefore, the Universe consisted of a plasma of nuclei, electrons, and photons. During this stage, the dominant components of matter were leptons, specifically electrons and neutrinos. Neutrinos are very light and weakly interacting particles that can pass through ordinary matter without obstruction. Neutrinos were formed through the beta decay of neutrons and also through the annihilation of electrons and positrons (anti-electrons). Intense annihilations between electrons and positrons occurred during the lepton epoch, resulting in the conversion of a significant portion of leptons into radiation. However, due to a violation of the symmetry between matter and antimatter during the electroweak interaction epoch, a small excess of electrons over positrons remained, with a ratio of approximately one electron for every ten billion antiparticles. This excess of electrons is what formed all the negatively charged matter in our Universe.

The lepton epoch is the subject of study in modern high-energy physics because it requires the theory of quantum electrodynamics for its description. According to this theory, electrons and positrons have electric charge and obey the law of charge conservation, meaning the total charge of a particle system remains unchanged during interactions. The electromagnetic interaction between electrons and positrons is mediated by the exchange of photons, which have no charge or mass. Photons also serve as quanta of light and other electromagnetic radiation. Neutrinos do not have electric charge but possess weak charge, which determines their interaction with other particles through the exchange of W and Z bosons. Neutrinos also have very small but non-zero mass, which can vary depending on the type of neutrino. There are three types of neutrinos: electron neutrino, muon neutrino, and tau neutrino, corresponding to the three types of leptons: electron, muon, and tau leptons. Neutrinos can undergo oscillation, a process where they can change from one type to another. This phenomenon has been experimentally confirmed and has shown that neutrinos have mass.

Photon Epoch

Following the lepton epoch, the photon epoch began, lasting for about 380,000 years. During this stage, the temperature of the Universe dropped to 3000 Kelvin, and the density decreased to 10^{-16} grams per cubic centimeter. As a result, the first atoms formed from nuclei and electrons through a process called recombination. Electrons combined with nuclei, forming neutral atoms. The first atoms were hydrogen and helium atoms, which constituted the majority of matter in the Universe. During the photon epoch, the dominant component of radiation was photons, the quanta of light and other electromagnetic radiation. Photons were produced through annihilations between particles and antiparticles and through transitions of electrons to lower energy levels in atoms. In the photon epoch, photons were released from matter, meaning they ceased interacting with atoms and freely propagated through the Universe. This phenomenon is called decoupling, and it made the Universe transparent to light.

The photon epoch is the subject of study in modern cosmology, as it requires the theory of cosmic microwave background radiation for its description. According to this theory, photons that were released from matter during the photon epoch have persisted to this day and constitute the cosmic microwave background radiation. This radiation has a blackbody spectrum with a temperature of approximately 2.7 Kelvin and serves as one of the main pieces of evidence for the Big Bang. It also contains information about the state of the Universe during the photon epoch, such as its density, temperature, expansion rate, and other parameters. By studying this radiation, we can learn a great deal about the early history and evolution of the Universe.

In this chapter, we have discussed the characteristics and stages of the Big Bang from its beginning to the formation of the first atoms and molecules. We have learned about the physical laws and processes that shaped the evolution of the Universe during different periods of its history, as well as the types of matter and radiation that dominated at each stage. We have also been introduced to some theoretical assumptions and hypotheses that attempt to explain certain enigmatic aspects of the Big Bang. In the next chapter, we will explore the traces and evidence of the Big Bang, such as cosmic microwave background radiation, the abundance of light elements, and the expansion of the Universe.

Traces and evidence of the Big Bang, such as cosmic microwave background radiation, abundance of light elements, and the expansion of the Universe.

In this chapter, we will examine the facts and observations that support the Big Bang theory and allow us to measure the parameters and characteristics of the Universe. We will learn about the radiation that remains from the early stages of the Universe's evolution and how it is distributed across the sky. We will also discover which elements were synthesized during the Big Bang and their abundance in space. Lastly, we will explore how the Universe expands over time and how it affects its geometry and fate.

Cosmic Microwave Background Radiation

One of the most important and compelling pieces of evidence for the Big Bang is the cosmic microwave background radiation (CMB), also known as relic radiation. This radiation is electromagnetic radiation with a wavelength of about 1 millimeter, filling all of space and having a temperature of approximately 2.7 Kelvin. It was discovered in 1965 by American physicists Arno Penzias and Robert Wilson, who were awarded the Nobel Prize in Physics in 1978 for this discovery.

The origin of the CMB is associated with the photon epoch, which occurred approximately 380,000 years after the Big Bang. During this stage, the temperature of the Universe dropped to 3000 Kelvin, allowing electrons to combine with nuclei and form neutral atoms. As a result, photons ceased interacting with matter and began freely propagating through the Universe. These photons have persisted to this day and constitute the CMB. However, over time, the Universe has expanded billions of times, causing spacetime to stretch, and the wavelength of photons to increase according to Hubble's law. As a result, photons that originally had a visible or infrared spectrum have transformed into microwave photons with lower energy and temperature.

The CMB has several features that allow us to study the state of the Universe during the photon epoch and its subsequent evolution. Firstly, the CMB exhibits a blackbody spectrum, meaning its intensity depends solely on temperature and is independent of direction or frequency of radiation. This indicates that the CMB was in thermodynamic equilibrium with matter during the photon epoch and was not subject to any external influences. Secondly, the CMB displays very small anisotropies, which are differences in the temperature of radiation in different directions. These anisotropies amount to about one-tenth of a thousandth of a degree and reflect tiny density fluctuations in the early Universe that served as the seeds of future galaxies and galaxy clusters. By studying these anisotropies, we can learn a great deal about the structure and geometry of the Universe, as well as its composition and parameters.

Special space observatories and telescopes have been developed to observe and measure the CMB, such as the Cosmic Background Explorer (COBE), the Wilkinson Microwave Anisotropy Probe (WMAP), and the Planck space telescope. These observatories have provided detailed maps of the CMB across the entire sky and have measured its spectrum and anisotropies with high precision and resolution. Thanks to this data, we can test and refine our model of the Big Bang and its predictions.

Abundance of light elements

Another important evidence for the Big Bang is the abundance of light elements, which refers to the ratio of light elements such as hydrogen, helium, lithium, and beryllium in space. These elements were synthesized during the Big Bang in a process called primordial or Big Bang nucleosynthesis. This process occurred approximately within the first 20 minutes after the Big Bang when the temperature of the Universe was high enough for nuclear reactions between protons and neutrons to take place. As a result of these reactions, the first atomic nuclei of light elements were formed.

The process of primordial nucleosynthesis depended on several factors, such as the density of baryonic matter (matter composed of protons and neutrons), the ratio of protons to neutrons, the rate of expansion of the Universe, and the neutrino decoupling time. By varying these factors, we can calculate the amount of light elements that should have been formed during the Big Bang. We can then compare these calculations with the observed abundance of light elements in different sources, such as stars, galaxies, interstellar, and intergalactic space. If these values match, it confirms the Big Bang theory and allows us to determine the parameters of the Universe.

However, in reality, comparing the calculated and observed abundance of light elements is not straightforward because there are many factors that can influence its variation. For example, nuclear reactions occur in stars, which can convert one element into another and also expel some matter into space in the form of stellar winds or supernova explosions. Additionally, processes such as mixing, diffusion, enrichment, or dilution of matter with different elements can occur in interstellar and intergalactic space. Therefore, to measure the primordial abundance of light elements, it is necessary to choose sources that are least affected by these factors. For example, the abundance of helium can be measured using the oldest and least evolved stars, while the abundance of deuterium (an isotope of hydrogen with one proton and one neutron) can be measured using the most distant and coldest gas clouds.

According to modern data, the abundance of light elements in space is approximately as follows: 75% hydrogen, 25% helium, and about 0.01% deuterium, lithium, and beryllium. These values are reasonably consistent with the calculations based on the Big Bang theory for certain values of baryonic matter density and the expansion rate of the Universe. However, there are some discrepancies and uncertainties that require further study and refinement. For example, the abundance of lithium in old stars is lower than predicted by the Big Bang theory, which could indicate that lithium is destroyed or lost in stars. The abundance of deuterium can also vary depending on the observation source, suggesting that deuterium is produced or destroyed in cosmic processes.

Expansion of the Universe

The third important evidence for the Big Bang is the expansion of the Universe, which refers to the increasing distances between galaxies over time. This phenomenon was discovered in 1929 by the American astronomer Edwin Hubble, who measured the recession velocities of galaxies based on the redshift of their spectral lines. He found that the farther a galaxy is from us, the faster it is receding, and that this relationship is linear. This means that there is a constant proportionality between velocity and distance, known as Hubble's constant. By measuring this constant, we can determine the rate of expansion of the Universe and its age.

The expansion of the Universe is explained by the Big Bang theory as a consequence of the stretching of spacetime that began with the Big Bang. According to this theory, the Universe is not expanding into any larger space; instead, spacetime itself is stretching and creating new regions. Thus, galaxies are not moving through space but rather moving away from each other due to the stretching of spacetime between them. This also means that the expansion of the Universe has no center or edge and occurs uniformly in all directions.

The expansion of the Universe depends on its density, temperature, velocity, and composition. The theory of general relativity by Albert Einstein is used to describe these dependencies, linking the geometry of spacetime with the distribution of matter and energy within it. According to this theory, the expansion of the Universe is described by the Friedman equations, which show how the scale factor—a parameter characterizing the size of the Universe—changes over time. From these equations, it follows that the Universe can have three possible scenarios of evolution: open, closed, or flat. An open Universe has insufficient matter and energy density to halt the expansion, so it will expand indefinitely and accelerate. A closed Universe has excess matter and energy density to halt the expansion, so it will initially expand and then contract towards a singularity in a process called the Big Crunch. A flat Universe has the critical matter and energy density to halt the expansion, so it will expand indefinitely and decelerate.

According to current data, our Universe is close to being flat and has a density of matter and energy of about 10^{-26} kilograms per cubic meter. However, the majority of this density is not accounted for by ordinary matter that we can see and measure, but by two mysterious components: dark matter and dark energy. Dark matter is an unknown form of matter that does not emit or absorb light but exerts gravitational influence on other objects. Dark energy is an unknown form of energy that permeates all of space and causes the acceleration of the Universe's expansion. According to modern estimates, dark matter makes up about 27% of the density of the Universe, dark energy accounts for about 68%, and ordinary matter comprises approximately 5%. The nature and origin of dark matter and dark energy remain among the biggest mysteries in modern cosmology.

Various methods and tools are used to measure the expansion of the Universe and its parameters, such as redshift, standard candles, gravitational lensing, and others. Redshift refers to the change in the wavelength of light emitted by distant objects due to the stretching of spacetime. The greater the redshift, the farther the object is from us and the faster it is receding. Standard candles are objects whose absolute brightness is known or can be calculated. By comparing the absolute and relative brightness of such objects, we can determine their distance from us. Examples of standard candles include Type Ia supernovae, which are explosions of white dwarfs in binary systems. Gravitational lensing is the distortion of the image of distant objects due to the bending of light in the gravitational field of nearby objects. By studying the shape and angular size of these distortions, we can determine the mass of the lensing object and the distance to both the lens and the lensed object.

Through these methods, we can construct the Hubble diagram, which shows the relationship between the recession velocity of galaxies and their distance from us. This diagram illustrates how the rate of the Universe's expansion changes over time and how it depends on the density of matter and energy within it. From this diagram, we can determine the Hubble constant, which is the proportionality coefficient between velocity and distance, as well as the deceleration parameter, which characterizes the deceleration or acceleration of the Universe's expansion. According to current data, the Hubble constant is approximately 70 kilometers per second per megaparsec, and the deceleration parameter is negative, indicating that the Universe is expanding with acceleration. From these values, we can calculate the age of the Universe, which is estimated to be about 13.8 billion years.

In this chapter, we have discussed the traces and evidence of the Big Bang that we can observe in our Universe and how they allow us to measure its parameters and characteristics. We have learned about the radiation that remains from the early stages of the Universe's evolution and how it is distributed across the sky. We have also explored the elements synthesized during the Big Bang and their abundance in space. Finally, we have examined how the Universe expands over time and how it affects its geometry and fate. In the next chapter, we will explore some of the mysteries and challenges of the Big Bang, such as dark matter, dark energy, antimatter, and the horizon problem.

Riddles and Problems of the Big Bang, such as dark matter, dark energy, antimatter, and the horizon problem.

In this chapter, we will explore the questions and challenges that arise when studying the Big Bang and its consequences. We will learn about the types of matter and energy that make up the majority of the Universe but still lack a satisfactory explanation. We will also discover the phenomena that violate the symmetry between matter and antimatter and between different regions of the Universe. Finally, we will explore the alternative and extended models of the Big Bang that attempt to address these problems and predict new effects.

Dark Matter

One of the biggest mysteries in modern cosmology is dark matter – an unknown form of matter that does not emit or absorb light but exerts gravitational influence on other objects. Dark matter constitutes about 27% of the density of the Universe and about 85% of all matter within it. The nature and origin of dark matter are still unknown.

The first hints of dark matter's existence emerged in the 1930s when Swiss astronomer Fritz Zwicky measured the velocities of galaxies in the Coma Cluster. He found that these velocities were too high for the cluster to be held together solely by its own gravity. He hypothesized the presence of additional invisible mass within the cluster, which provided galaxies with extra speed and stabilized the cluster. He named this invisible mass dark matter.

Subsequent observations were made for other galaxy clusters as well as individual galaxies. It was discovered that dark matter is distributed unevenly throughout the Universe, forming gigantic halos around galaxies and galaxy clusters. It was also found that dark matter plays a crucial role in the formation of the Universe's structure, as it serves as the scaffold for the formation of ordinary matter in the form of stars and galaxies. Without dark matter, the Universe would appear vastly different.

There are many hypotheses about what dark matter could be. One of the most popular is the hypothesis of cold dark matter (CDM), which posits that dark matter consists of slow-moving and massive particles that interact only through gravity. These particles could be new elementary particles that are not part of the standard model of physics, such as neutralinos or axions. Another hypothesis is the warm dark matter (WDM) hypothesis, which suggests that dark matter consists of fast and light particles, such as sterile neutrinos. Yet another hypothesis is the modified gravity (MOG) hypothesis, which proposes that dark matter does not exist at all, and its effects are explained by modifications to the laws of gravity on large scales.

Various methods and experiments are employed to test these hypotheses and search for dark matter. These include observations of gravitational lensing, measurements of cosmic microwave background anisotropies, the search for dark matter particles in cosmic rays or ground-based detectors, the production of dark matter particles in particle accelerators, and more. However, none of these methods have provided a definitive answer to the question of what dark matter is and where it comes from.

Dark Energy

Another major mystery in modern cosmology is dark energy – an unknown form of energy that permeates all of space and drives the accelerated expansion of the Universe. Dark energy constitutes about 68% of the density of the Universe and about 95% of all energy within it. The nature and origin of dark energy are still unknown.

The first hints of dark energy's existence emerged in 1998 when two independent groups of astronomers measured the distances to Type Ia supernovae in distant galaxies. They found that these distances were greater than what the Big Bang theory predicted without dark energy. This indicated that the Universe was expanding with acceleration rather than slowing down, as expected. To explain this phenomenon, they hypothesized the existence of additional energy in the Universe that acts as anti-gravity and counteracts the gravity of ordinary matter. They named this additional energy dark energy.

Subsequent observations confirmed the existence of dark energy using various methods and instruments, such as measurements of cosmic microwave background anisotropies, measurements of baryon acoustic oscillations in the distribution of galaxies, measurements of structure growth in space, and others. All these data demonstrate that dark energy does indeed exist and has a significant influence on the evolution of the Universe.

There are many hypotheses about what dark energy could be. One of the most popular is the hypothesis of the cosmological constant, which suggests that dark energy is an inherent property of space-time itself, remaining constant over time and independent of location. This constant was introduced by Albert Einstein in his equations of general relativity to make the Universe static and stable. However, he later abandoned this idea, considering it his biggest mistake. However, modern observations indicate that the cosmological constant may not be zero but have a very small positive value, corresponding to the observed dark energy.

Another hypothesis is the quintessence hypothesis, which suggests that dark energy is a dynamic field that changes over time and depends on location. Such a field could be analogous to the Higgs field, which gives elementary particles their mass, or the inflaton field, which is responsible for the inflation of the Universe. Such a field can have different properties and behaviors depending on its potential and equation of state. For example, it can be constant or variable, homogeneous or inhomogeneous, stable or unstable, and so on.

Another hypothesis is the modified gravity (MOG) hypothesis, which proposes that dark energy does not exist at all, and its effects are explained by modifications to the laws of gravity on large scales. Such modifications could be caused by additional dimensions of space-time, nonlinearity of the Einstein equations, violation of Lorentz symmetry, and other factors.

Various methods and experiments are employed to test these hypotheses and measure the parameters of dark energy. These include observations of Type Ia supernovae, measurements of cosmic microwave background anisotropies, measurements of baryon acoustic oscillations in the distribution of galaxies, measurements of structure growth in space, and others. However, none of these methods have provided a definitive answer to the question of what dark energy is and where it comes from.

Antimatter

Another major mystery in modern cosmology is antimatter – a form of matter that consists of antiparticles, which have opposite charge and other quantum numbers compared to ordinary particles. Antimatter annihilates upon contact with ordinary matter, releasing a tremendous amount of energy. The nature and origin of antimatter are still unknown.

The first antiparticle, the positron (anti-electron), was discovered in 1932 by American physicist Carl Anderson, who received the Nobel Prize in Physics in 1936 for this discovery. Subsequent discoveries included other antiparticles such as antiprotons, antineutrons, antinucleons, and others. Anti-atoms, such as antihydrogen and anti-helium, have also been created and studied. However, the amount of antimatter in our Universe is extremely small compared to ordinary matter. Estimates suggest that for every billion particles of matter, there is only one antiparticle. This is one of the greatest asymmetries in nature.

The origin of this asymmetry is related to the electroweak epoch, which occurred approximately within the first 10^{-12} seconds after the Big Bang. During this stage, the temperature of the Universe was around 10^{15} Kelvin, and the density was about 10^{25} grams per cubic centimeter. Intense reactions took place between particles and antiparticles, where they were being created and annihilated. According to the Big Bang theory, during this epoch, there should have been an equal number of particles and antiparticles, as the laws of physics were expected to be symmetric with respect to matter and antimatter. However, observations show that this is not the case, and that matter dominates over antimatter in our Universe.

To explain this phenomenon, it is necessary to hypothesize that there was a violation of symmetry between matter and antimatter during the electroweak epoch, meaning that some processes were more likely to occur for particles than for antiparticles, or vice versa. This violation of symmetry is called baryogenesis or baryon asymmetry, and it means that a small excess of particles over antiparticles was formed as a result of reactions between particles and antiparticles. This excess amounted to about one particle per billion antiparticles. When the temperature of the Universe dropped to a level where reactions between particles and antiparticles ceased, this excess remained and became the basis for the formation of all matter in our Universe.

To enable baryogenesis, three conditions formulated by the Soviet physicist Andrei Sakharov in 1967 must be satisfied. These conditions are as follows:

Violation of baryon number: This means that processes must exist in which the number of baryons (particles composed of three quarks, such as protons and neutrons) can change. For example, in such processes, baryons can be created or destroyed, or they can transform into other particles.

Violation of C and CP symmetry: This means that processes must exist in which particles and antiparticles behave differently. C symmetry is the symmetry with respect to the interchange of particles and antiparticles. CP symmetry is the symmetry with respect to the interchange of particles and antiparticles combined with a change in the sign of spatial coordinates. For example, in such processes, the laws of charge conservation or parity may be violated.

Lack of thermodynamic equilibrium: This means that processes must exist in which the rates of reactions between particles and antiparticles are not equal. For example, irreversible or non-equilibrium factors, such as the expansion of the Universe or interactions with other fields, can be involved in such processes.

There are many theoretical models attempting to explain baryogenesis and the fulfillment of these conditions. One of the most popular models is the electroweak baryogenesis model, which suggests that baryogenesis occurred during the phase transition from electroweak interactions to electromagnetic and weak interactions. During this transition, bubbles of a new phase of spacetime could have formed, expanded, and collided with each other. As a result of these collisions, irreversible and non-equilibrium processes could have taken place, leading to violations of C and CP symmetries and changes in the baryon number. However, this model has several problems and inconsistencies with observations, thus requiring further development and modification.

To test these models and search for antimatter, various methods and experiments are employed, such as observations of matter-antimatter annihilation in cosmic rays or on ground-based detectors, creation of antimatter in particle accelerators, search for antigalaxies or anti-regions in space, and others. However, none of these methods have provided a definitive answer to why matter prevails over antimatter in our Universe.

The Horizon Problem

Another major problem in modern cosmology is the horizon problem - the problem of explaining the homogeneity and isotropy of the Universe on large scales. Homogeneity means that the density and temperature of the Universe do not depend on location, while isotropy means that they do not depend on direction. Observations show that the Universe is indeed homogeneous and isotropic to an accuracy of 0.01%, as confirmed, for example, by the uniformity of cosmic microwave background radiation across the entire sky.

However, the Big Bang theory cannot explain why the Universe is so homogeneous and isotropic. The problem arises from the fact that different regions of the Universe could not have been in thermal equilibrium with each other during the Big Bang because they were too far apart and could not exchange information or energy. Such regions are called causally disconnected or lying beyond each other's horizons. The size of such a horizon depends on time and the rate of expansion of the Universe. According to the Big Bang theory, the size of the horizon during the photon epoch was about 2 degrees on the celestial sphere. This means that different regions of the sky that are separated by more than 2 degrees could not interact with each other during the Big Bang. So how can these regions have the same temperature and density?

To solve this problem, the hypothesis of cosmic inflation was proposed, which suggests that in the very beginning of the universe's evolution, there was a rapid and strong growth of its size - inflation. During inflation, the universe expanded by billions upon billions of times in a fraction of a second. This allowed for the smoothing out of all the inhomogeneities and anisotropies in the universe, making it nearly perfectly homogeneous and isotropic. It also enlarged the horizon to such an extent that all the regions of the sky we observe now were causally connected to each other before inflation. Thus, inflation resolves the horizon problem and explains the uniformity and isotropy of the universe.

There are many theoretical models attempting to explain inflation and its mechanism. One of the most popular models is the chaotic inflation model, which suggests that inflation is caused by the dynamics of a scalar field called the inflaton. This field has a very high energy potential that dominates over all other forms of energy in the universe. When the potential reaches a certain value, the field slowly rolls down towards its minimum, causing spacetime to exponentially stretch. When the field reaches the minimum of its potential, it oscillates around it and releases its energy in the form of other particles and fields. This process is called reheating and marks the end of inflation and the beginning of the universe's ordinary evolution.

Various methods and experiments are employed to test these models and search for traces of inflation, such as measurements of cosmic microwave background anisotropies, measurements of primordial gravitational waves, measurements of large-scale structure statistics in the universe, and others. However, none of these methods have provided a definitive answer yet regarding whether inflation occurred and how it unfolded.

In this chapter, we have discussed the mysteries and problems that arise when studying the Big Bang and its consequences. We have learned about the types of matter and energy that constitute the majority of the universe but still lack a satisfactory explanation. We have also explored the phenomena that break symmetry between matter and antimatter and between different regions of the universe. Finally, we have explored alternative and extended models of the Big Bang that attempt to address these issues and predict new effects. In the next chapter, we will delve into some of these models in more detail and explore the prospects for further research.

Alternative and extended models of the Big Bang, such as the multiverse, cyclic universe, quantum gravity, and string theory.

In this chapter, we will explore the other possible scenarios for the birth and development of the universe beyond the standard model of the Big Bang. We will learn about the new physical principles and hypotheses underlying these models and the new phenomena and effects they predict. We will also discover the experimental data and observations that can confirm or refute these models and the problems and challenges they face.

The Multiverse

One of the most radical and controversial alternative models of the Big Bang is the hypothesis of the multiverse, which suggests that our universe is not the only one but rather one of many other universes that exist in parallel or sequentially with each other. These universes may have different properties, laws of physics, constants of nature, sizes, shapes, and histories. The nature and origin of the multiverse are still unknown.

There are many different approaches and classifications to the concept of the multiverse. One of the most popular is the so-called Tegmark scheme, which divides the multiverse into four levels based on the degree of difference between universes. These levels are as follows:

Level 1: Quasi-Classical Multiverse. This is a multiverse in which all universes have the same laws of physics and constants of nature but differ in initial conditions and random fluctuations. Such universes arise from the infinite expansion of our universe and its regions beyond the observational horizon. For example, these universes may have more or fewer galaxies, stars, or planets than our universe.

Level 2: Inflationary Multiverse. This is a multiverse in which all universes have the same fundamental laws of physics but differ in the values of the constants of nature and symmetry parameters. Such universes arise from eternal inflation, a process in which certain regions of spacetime continue to undergo exponential expansion while others transition to the ordinary phase. For example, these universes may have different sizes, dimensions, forces, or particles than our universe.

Level 3: Quantum Multiverse. This is a multiverse in which all universes have the same laws of physics and constants of nature but differ in history and the outcomes of quantum measurements. Such universes arise from decoherence, a process in which the quantum states of a system lose coherence due to interactions with the surrounding environment. For example, these universes may have different outcomes for the same experiment or event than our universe.

Level 4: Platonic Multiverse. This is a multiverse in which all universes have different laws of physics, constants of nature, as well as different mathematical structures and logics. Such universes arise from the realization of all possible mathematical equations and abstractions. For example, these universes may have completely different geometries, algebras, arithmetic, or logics than our universe.

There are many different arguments and reasons for or against the existence of a multiverse. One of the strongest arguments in favor of a multiverse is the anthropic principle, which states that we observe our universe as it is because only in such a universe could life and intelligence arise and evolve. In other words, we are the product of a random selection from a set of possible universes. This principle explains why our universe has finely-tuned parameters and constants that make it suitable for life. For example, if the force of gravity were slightly stronger or weaker, stars could not form or would quickly burn out. If the mass ratio of a proton to an electron were slightly different, atoms could not exist or would be unstable. If the energy density of the universe were slightly higher or lower, the universe would quickly collapse or disintegrate. Thus, the anthropic principle avoids the problem of fine-tuning and makes our universe not special but just one among many.

However, there are other arguments and reasons against the existence of a multiverse. One of the strongest arguments against a multiverse is the problem of falsifiability, which states that we cannot test or refute the hypothesis of a multiverse because we cannot observe or interact with other universes. In other words, the multiverse is not a scientific theory but merely a philosophical speculation. This argument calls into question the value and meaning of reasoning about the multiverse and its properties. Additionally, this argument contradicts Occam's razor, which states that entities should not be multiplied unnecessarily. In other words, we should not assume the existence of a multiverse if we can explain our universe without it.

Various methods and experiments are used to test the hypothesis of a multiverse and search for its traces, such as measurements of cosmic microwave background anisotropies, measurements of primordial gravitational waves, the search for collisions or overlaps between universes, and others. However, none of these methods has provided a definitive answer to the question of whether a multiverse exists and what it is like.

The Cyclical Universe

Another alternative model to the Big Bang is the hypothesis of a cyclical universe, which suggests that our universe is not the first and only one but rather one of an infinite number of universes that arise and disappear through periodic processes of contraction and expansion. These processes can be caused by various physical mechanisms, such as fluctuations in the energy density of the universe, interactions with other universes, or other dimensions of space-time. The nature and origin of the cyclical universe are still unknown.

There are many different approaches and models concerning the concept of a cyclical universe. One of the most popular is the model of an ekpyrotic cyclic universe (ECU), which assumes that our universe undergoes an infinite sequence of expansion and contraction phases, each having the same duration and amplitude. In each cycle, the universe goes through four stages: inflation, radiation, matter, and vacuum. During the inflation stage, the universe undergoes exponential expansion, smoothing out all inhomogeneities and anisotropies. In the radiation stage, the universe is filled with warm radiation that dominates over all other forms of energy. In the matter stage, the universe is filled with cold matter that dominates over all other forms of energy. In the vacuum stage, the universe is filled with dark energy that dominates over all other forms of energy and causes the acceleration of the universe's expansion. At the end of this stage, the universe reaches its maximum size and begins to contract under the influence of its own gravity. This leads to a reverse inflation, during which the universe contracts exponentially, increasing its density and temperature. This process culminates in a Big Crunch, where the universe reaches its minimum size and maximum density and temperature. After that, a new cycle begins with a new inflation.

Various methods and experiments are used to test the hypothesis of a cyclical universe and search for its traces, such as measurements of cosmic microwave background anisotropies, measurements of primordial gravitational waves, the search for anomalies in the large-scale structure of the universe, and others. However, none of these methods has provided a definitive answer to the question of whether a cyclical universe exists and what it is like.

Quantum Gravity

Another alternative model to the Big Bang is the hypothesis of quantum gravity, which suggests that our universe arose from quantum fluctuations of spacetime at the Planck scale. The Planck scale is the smallest and most energetic scale in physics, where all four fundamental forces of nature—gravity, electromagnetism, strong, and weak nuclear interactions—operate. At this scale, spacetime ceases to be smooth and continuous and becomes discrete and uncertain. The nature and origin of quantum gravity are still unknown.

There are many different approaches and models to the concept of quantum gravity. One of the most popular is string theory, which posits that all particles and fields in nature are manifestations of a single unified object called a string, which can vibrate in different modes and dimensions. A string is a one-dimensional object with a length on the order of the Planck length (about 10^{-35} meters) and zero thickness. The string can exist in different types of spacetime that have varying numbers of dimensions (from 10 to 26). The string can move and interact with other strings, forming complex structures and configurations. It can also loop back on itself, forming a loop or ring. The string can break or join with other strings, altering its topology.

String theory allows for the unification of all four fundamental forces of nature into a single unified theory. In particular, string theory enables the quantization of gravity and describes it as one of the vibration modes of the string. Thus, string theory resolves the problem of the incompatibility between general relativity and quantum mechanics that arises at the Planck scale. It also explains the origin and properties of all elementary particles and fields in nature, as they are manifestations of different vibration modes of the string.

Various methods and experiments are used to test the hypothesis of quantum gravity and search for its traces, such as measurements of cosmic microwave background anisotropies, measurements of primordial gravitational waves, the search for minimal length or spacetime discreteness, and others. However, none of these methods has provided a definitive answer to the question of whether quantum gravity exists and what it is like.

In this chapter, we have explored some of the alternative and extended models of the Big Bang that attempt to explain the origin and evolution of the universe by considering new physical principles and hypotheses. We have learned about the new phenomena and effects predicted by these models and the new challenges and difficulties they encounter. We have also discussed the experimental data and observations that can confirm or refute these models and the prospects for further research.

In the next chapter, we will summarize our discussion of the Big Bang and the birth of the universe. We will consolidate what we know and do not know about this process and what questions and mysteries remain open. We will also discuss the current directions and methods of cosmological development and the expectations and hopes associated with them.

What we know and don't know about the Big Bang and the birth of the Universe, and the prospects for further research.

In this book, we have discussed how modern science is trying to understand the origin and development of the Universe through the theory of the Big Bang. We have learned about the historical and cultural factors that influenced the formation of this theory and the experimental data and observations that support it. We have also learned about the key characteristics and stages of the Big Bang and the traces and evidence it has left in our Universe. Finally, we have learned about the puzzles and problems that arise when studying the Big Bang and the alternative and expanded models that attempt to solve them.

We have found that the theory of the Big Bang is one of the most successful and convincing theories in physics and cosmology. It explains many observed facts and phenomena, such as the expansion of the Universe, cosmic microwave background radiation, the abundance of light elements, the formation of galaxies and stars, and more. It also predicts many new effects and phenomena that can be tested experimentally or observed. It is also consistent with the fundamental principles and laws of physics and logic.

However, we have also found that the theory of the Big Bang is not a complete and final theory. It has several problems and inconsistencies that require further development and modification. It cannot explain the origin of the Big Bang itself and the singularity that preceded it. It cannot explain the nature and origin of dark matter and dark energy, which make up a large part of the content of the Universe. It cannot explain the violation of symmetry between matter and antimatter and between different regions of the Universe. It cannot explain the fine-tuning of the parameters and constants of nature that make our Universe suitable for life. It cannot explain the existence of life and consciousness in our Universe.

To solve these problems, there are various alternative and expanded models of the Big Bang that try to explain the origin and development of the Universe by taking into account new physical principles and hypotheses. We have discussed some of these models, such as the multiverse, cyclic Universe, quantum gravity, and string theory. We have learned about the new phenomena and effects predicted by these models and the new problems and difficulties they have. We have also learned about the experimental data and observations that can confirm or refute these models.

We have concluded that there is currently no unified and final theory that explains all aspects of the Big Bang and the birth of the Universe. Science continues to search for answers to these questions and develop new methods and tools for their study. Science also continues to encounter new mysteries and problems that require new hypotheses and models. Science also continues to be amazed by the beauty and complexity of our Universe and its laws.

We hope that this book has helped you understand what the Big Bang is and why it is important to understand the origin of the Universe. We also hope that this book has sparked your interest in cosmology and physics and encouraged you to further study these sciences. We also hope that this book has shown you that science is not just dry facts and formulas but also fascinating stories and adventures that open up new horizons of knowledge and understanding. We also hope that this book has prompted you to think about what our Universe really is and what our place in it is.

Thank you for your attention!

www.ingramcontent.com/pod-product-compliance
Lightning Source LLC
Chambersburg PA
CBHW070137230526
45472CB00004B/1576